MW00513430

# CIHM
# Microfiche
# Series
# (Monographs)

# ICMH
# Collection de
# microfiches
# (monographies)

Canadian Institute for Historical Microreproductions / Institut canadien de microreproductions historiques

©1995

Technical and Bibliographic Notes / Notes techniques et bibliographiques

The Institute has attempted to obtain the best original copy available for filming. Features of this copy which may be bibliographically unique, which may alter any of the images in the reproduction, or which may significantly change the usual method of filming, are checked below.

L'Institut a microfilmé le meilleur exemplaire qu'il lui a été possible de se procurer. Les détails de cet exemplaire qui sont peut-être uniques du point de vue bibliographique, qui peuvent modifier une image reproduite, ou qui peuvent exiger une modification dans la méthode normale de filmage sont indiqués ci-dessous.

☑ Coloured covers/
Couverture de couleur

☐ Covers damaged/
Couverture endommagée

☐ Covers restored and/or laminated/
Couverture restaurée et/ou pelliculée

☐ Cover title missing/
Le titre de couverture manque

☐ Coloured maps/
Cartes géographiques en couleur

☑ Coloured ink (i.e. other than blue or black)/
Encre de couleur (i.e. autre que bleue ou noire)

☑ Coloured plates and/or illustrations/
Planches et/ou illustrations en couleur

☐ Bound with other material/
Relié avec d'autres documents

☐ Tight binding may cause shadows or distortion along interior margin/
La reliure serrée peut causer de l'ombre ou de la distorsion le long de la marge intérieure

☐ Blank leaves added during restoration may appear within the text. Whenever possible, these have been omitted from filming/
Il se peut que certaines pages blanches ajoutées lors d'une restauration apparaissent dans le texte, mais, lorsque cela était possible, ces pages n'ont pas été filmées.

☐ Coloured pages/
Pages de couleur

☐ Pages damaged/
Pages endommagées

☐ Pages restored and/or laminated/
Pages restaurées et/ou pelliculées

☑ Pages discoloured, stained or foxed/
Pages décolorées, tachetées ou piquées

☐ Pages detached/
Pages détachées

☑ Showthrough/
Transparence

☐ Quality of print varies/
Qualité inégale de l'impression

☐ Continuous pagination/
Pagination continue

☐ Includes index(es)/
Comprend un (des) index

Title on header taken from:/
Le titre de l'en-tête provient:

☐ Title page of issue/
Page de titre de la livraison

☐ Caption of issue/
Titre de départ de la livraison

☐ Masthead/
Générique (périodiques) de la livraison

☐ Additional comments:/
Commentaires supplémentaires:

This item is filmed at the reduction ratio checked below/
Ce document est filmé au taux de réduction indiqué ci-dessous.

| 10X | | 14X | | 18X | | 22X | | 26X | | 30X | |
|---|---|---|---|---|---|---|---|---|---|---|---|
| | 12X | | 16X ✓ | | 20X | | 24X | | 28X | | 32X |

The copy filmed here has been reproduced thanks to the generosity of:

National Library of Canada

The images appearing here are the best quality possible considering the condition and legibility of the original copy and in keeping with the filming contract specifications.

Original copies in printed paper covers are filmed beginning with the front cover and ending on the last page with a printed or illustrated impression, or the back cover when appropriate. All other original copies are filmed beginning on the first page with a printed or illustrated impression, and ending on the last page with a printed or illustrated impression.

The last recorded frame on each microfiche shall contain the symbol ➔ (meaning "CONTINUED"), or the symbol ▽ (meaning "END"), whichever applies.

Maps, plates, charts, etc., may be filmed at different reduction ratios. Those too large to be entirely included in one exposure are filmed beginning in the upper left hand corner, left to right and top to bottom, as many frames as required. The following diagrams illustrate the method:

L'exemplaire filmé fut reproduit grâce à la générosité de:

Bibliothèque nationale du Canada

Les images suivantes ont été reproduites avec le plus grand soin, compte tenu de la condition et de la netteté de l'exemplaire filmé, et en conformité avec les conditions du contrat de filmage.

Les exemplaires originaux dont la couverture en papier est imprimée sont filmés en commençant par le premier plat et en terminant soit par la dernière page qui comporte une empreinte d'impression ou d'illustration, soit par le second plat, selon le cas. Tous les autres exemplaires originaux sont filmés en commençant par la première page qui comporte une empreinte d'impression ou d'illustration et en terminant par la dernière page qui comporte une telle empreinte.

Un des symboles suivants apparaîtra sur la dernière image de chaque microfiche, selon le cas: le symbole ➔ signifie "A SUIVRE", le symbole ▽ signifie "FIN".

Les cartes, planches, tableaux, etc., peuvent être filmés à des taux de réduction différents. Lorsque le document est trop grand pour être reproduit en un seul cliché, il est filmé à partir de l'angle supérieur gauche, de gauche à droite, et de haut en bas, en prenant le nombre d'images nécessaire. Les diagrammes suivants illustrent la méthode.

| 1 | 2 | 3 |
|---|---|---|

| 1 |
|---|
| 2 |
| 3 |

| 1 | 2 | 3 |
|---|---|---|
| 4 | 5 | 6 |

# MICROCOPY RESOLUTION TEST CHART

(ANSI and ISO TEST CHART No. 2)

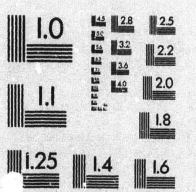

 APPLIED IMAGE Inc

1653 East Main Street
Rochester, New York 14609 USA
(716) 482 - 0300 - Phone
(716) 288 - 5989 - Fax

# ALLAN LINE

## & State Line

### STEAMSHIPS.

# INFORMATION FOR PASSENGERS.

# ALLAN LINE

## ROYAL MAIL STEAMSHIPS.

---

## INFORMATION FOR PASSENGERS.

---

### GENERAL WESTERN AGENTS.

ALLAN & CO. - - - - - - CHICAGO.

H. BOURLIER, - - - - - - TORONTO.

R. KERR, - - - - - - - WINNIPEG.

---

## H. & A. ALLAN,

### GENERAL AGENTS.

STATE STREET, - - - BOSTON.

225 WALNUT STREET, PHILADELPHIA.

25 COMMON STREET, - MONTREAL.

MONTREAL, April 2nd, 1892.

# ALLAN LINE SERVICES.

## LIVERPOOL, QUEBEC & MONTREAL.
### (in Summer),
### Via LONDONDERRY.

## LIVERPOOL AND PORTLAND.
### (in Winter),
### Via LONDONDERRY and HALIFAX.

## LIVERPOOL AND BALTIMORE.
### Via QUEENSTOWN,
### Calling at HALIFAX, N.S., and ST. JOHNS, N.F.

## GLASGOW AND NEW YORK.
### Via LONDONDERRY.

## GLASGOW AND BOSTON.
### Via LONDONDERRY and GALWAY.

## GLASGOW AND PHILADELPHIA.
### Via LONDONDERRY and GALWAY.

## GLASGOW, QUEBEC & MONTREAL.

## LONDON, QUEBEC & MONTREAL.

## GLASGOW, LIVERPOOL & RIVER PLATE.

# ALLAN · LINE.

This Company's Lines are composed of the following
Double Engined Clyde Built

## IRON AND STEEL STEAMSHIPS.

They are built in water-tight compartments, are unsurpassed for strength, speed and comfort, are fitted up with all the modern improvements that practical experience can suggest, and have MADE SOME OF THE FASTEST TRIPS ON RECORD.

|  | TONNAGE. | | |
|---|---|---|---|
| ACADIAN | 931 | Capt. | C. Mylius. |
| ASSYRIAN | 3970 | Capt. | John Bentley. |
| AUSTRIAN | 2458 | Capt. | Fairfull. |
| BRAZILIAN | 4100 | Capt. | Whyte. |
| BUENOS AYREAN | 4005 | Capt. | Vipond. |
| CANADIAN | 2906 | Capt. | J. Goodwin. |
| CARTHAGINIAN | 4214 | Capt. | John France. |
| CASPIAN | 2728 | Capt. | John Wallace. |
| CIRCASSIAN | 3724 | Capt. | R. P. Moore. |
| COREAN | 3488 | Capt. | C. J. Menzies. |
| GRECIAN | 3613 | Capt. | C. E. Le Gallais. |
| HIBERNIAN | 2997 | Capt. | Wylie. |
| LUCERNE | 1925 | Capt. | Stewart. |
| MANITOBAN | 2975 | Capt. | McAdam. |
| MONGOLIAN | 4750 | Lieut. | R. Barrett, R.N.R. |
| MONTE VIDEAN | 3500 | Capt. | A. Ferguson. |
| NESTORIAN | 2689 | Capt. | Gunsen. |
| NORWEGIAN | 3523 | Capt. | W. Christie. |
| NOVA SCOTIAN | 3305 | Capt. | R. H. Hughes. |
| NUMIDIAN | 4750 | Capt. | A. Macnicol. |
| PARISIAN | 5359 | Capt. | Joseph Ritchie. |
| PERUVIAN | 3038 | Capt. | Nunan. |
| PHŒNICIAN | 2425 | Capt. | D. J. James. |
| POLYNESIAN | 3983 | Capt. | Alex. McDougall. |
| POMERANIAN | 4364 | Capt. | W. Dalziel. |
| PRUSSIAN | 3080 | Capt. | Calvert. |
| ROSARIAN | 3500 | Capt. | Dunlop. |
| SARDINIAN | 4376 | Capt. | Wm. Richardson. |
| SARMATIAN | 3647 | Capt. | Johnstone. |
| SCANDINAVIAN | 3068 | Capt. | Stirrat. |
| SIBERIAN | 3904 | Capt. | John Park. |
| WALDENSIAN | 2256 | Capt. | Brodie. |
| STATE OF CALIFORNIA | 5500 | Capt. | Braes. |
| STATE OF NEBRASKA | 4000 | Capt. | John Brown. |
| STATE OF NEVADA | 3000 | Capt. | Main |
| STATE OF PENNSYLVANIA | 3000 | Capt. | McCulloch. |

# ALLAN LINE

## ROYAL MAIL STEAMSHIPS.

**The Allan Line** is one of the oldest in existence, having been uninterruptedly carrying passengers across the Atlantic for nearly seventy years. So far back as 1820, Allan Line sailing ships went regularly to Canada, and the excellent character and build of these vessels, the carefulness with which they were managed, and the attention and care bestowed upon passengers, soon made them well known and popular among ocean travellers. New vessels were constantly added, every improvement that scientific knowledge and experience suggested was unhesitatingly adopted, and a great passenger trade was carried on by Allan sailing ships for about fifty years, these vessels continuing to carry passengers many years after steamers commenced to run.

In 1852 the Canadian Government alive to the importance of direct Steam communication between its provinces and Britain, determined that a weekly Steam line should be established. Sir Hugh Allan and Mr. Andrew Allan of Montreal, on behalf of the Allan Line, tendered for this service. Their offer was accepted and the Canadian Mail Line of Steamers was inaugurated. The Steamers built by the Messrs. Allan at a bound became favorites not only with the Canadian community, but with general travellers.

Larger steamers have been added to the fleet from time to time, each new vessel having all the improvements that new inventions or discoveries have shown to be valuable to ensure safety, speed and comfort. A long experience, longer than that of any other Company now carrying passengers across the Atlantic, enriched by the practical observation that did not overlook even minute details, resulted in a fleet of steamers specially adapted to passenger trade.

Many persons accustomed to sail out of New York, now prefer the staunch and comfortable Steamers of the Allan Line from Montreal and Quebec via the pleasant St. Lawrence route, and the cabin, second cabin and steerage traffic, both to and from Europe, is largely increased from this source.

In the face of many discouragements, the Allan Line has always endeavoured to be foremost in everything tending to the advancement of Canada, its people and its commerce—and this policy will be energetically maintained.

### DESCRIPTION OF

## The Allan Line Royal Mail Steamship

# "PARISIAN."

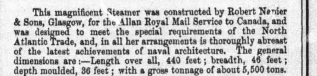

This magnificent Steamer was constructed by Robert Napier & Sons, Glasgow, for the Allan Royal Mail Service to Canada, and was designed to meet the special requirements of the North Atlantic Trade, and, in all her arrangements is thoroughly abreast of the latest achievements of naval architecture. The general dimensions are:—Length over all, 440 feet; breadth, 46 feet; depth moulded, 36 feet; with a gross tonnage of about 5,500 tons.

The machinery is capable of developing 6,000 indicated horse power and of propelling the Ship 16 knots per hour. To obtain the maximum of strength the hull has been built of steel, and the bottom has been constructed with an inner and outer skin, five feet apart; this space is also sub-divided into numerous water-tight compartments, and while available for water ballast, is principally valuable should the outer skin be damaged, while the water-tight bulkheads protect the vessel from the perils of collision. Mess.s. Allan were the first to apply this style of build to large Atlantic Steamers, and also the first to build such steamers of steel.

The Parisian is fitted for 160 Cabin Passengers in the most complete and splendid style; every modern improvement and convenience being introduced. She can also carry 120 Second Cabin, and 1,000 Steerage passengers. By a new and admirable arrangement of side keels, she has been made so wonderfully steady at sea, that sea sickness on board her is practically unknown.

Her Royal Highness The Princess Louise, in her voyages to and from Canada, has invariably selected Allan Line Steamers,— a convincing proof of her belief in their seaworthiness, comfort and management.

**THE "PARISIAN" IS LIGHTED THROUGHOUT BY ELECTRICITY.**

B

# THE ST. LAWRENCE ROUTE
## TO EUROPE.
### 480 miles less than from New York.

MONTREAL, AS A PORT OF EMBARKATION for the tourist or traveller *en route* to Europe, has unrivalled advantages, making it pre-eminent over any port on the American continent from which to commence a trip either for pleasure or business.

The distance to the seaboard, and facilities for reaching it, should be the first consideration of the traveller, and in this respect, MONTREAL is far more advantageously situate than its American Rival New York. Taking Chicago as an example, the distance from that City to Montreal is about 150 miles less than to New York, and instead of a long journey by rail, the trip may be diversified by a run down to Niagara Falls or TORONTO, taking from the latter City one of the splendid Steamers of the Richelieu and Ontario Navigation Company, through Lake Ontario and down the St. Lawrence, through the far famed Thousand Islands, arriving at Montreal in time to spend a few hours before embarking —a trip in itself far superior to many made by the ordinary tourist during his visit in Europe.

When the steamer leaves Montreal the voyage commences, but the uncomfortable sensation which so many Ocean travellers look forward to with dread is changed to the pleasurable one of "sailing on a summer sea."

The grand old historic City of Quebec is reached early in the afternoon, giving passengers an opportunity of spending a few hours on shore, the steamer not sailing until 9 o'clock the following morning.

Space will not admit of a full description of the grand and attractive scenery of the River and Gulf of the St. Lawrence, nearly a thousand miles of which is traversed after leaving Quebec, it is sufficient to say that when after about three days the Ocean is reached, it is not surprising that the passengers are in the best of spirits, and that the novelty of being aboard ship has been forgotten by them in the feeling of satisfaction, that nothing has been left undone by the Company that could tend to their security and comfort.

# ITINERARY OF SERVICES.

## LIVERPOOL.

Steamers sail from this port regularly every week during the Summer months for Quebec and Montreal, calling at Londonderry, and in the Winter regularly to Portland, calling at Londonderry and Halifax. There is also a fortnightly service to Baltimore, calling at Queenstown, St. Johns, N.F., and Halifax. Passengers may be booked from Liverpool via the Glasgow Services to Boston or Philadelphia.

## GLASGOW.

From this port there is a weekly Service to New York and a fortnightly service to Boston and to Philadelphia, together with a weekly Service during St. Lawrence navigation, to Quebec and Montreal. Passengers from Scotla⁷ therefore, have a choice of a direct route to either the New E⁷and, Middle, Southern and Western States, or Canada.

## LONDON.

The Service from this port is a Direct one to Quebec and Montreal, and only during the Summer months.

## BRISTOL AND CARDIFF.

Prepaid passengers booked from these ports to points in the United States, are furnished with transportation to Liverpool or Glasgow, and thence by the Service most convenient to the destination of the passenger. This also applies to Prepaid passengers from London.

## LONDONDERRY, (MOVILLE.)

Is situate in the North of Ireland, and is consequently most direct for passengers from that section and Middle of Ireland, as it saves them the trip by water to Liverpool, (which they must take if they sail from that port) and the consequent delays attending transfer. Allan Line Passengers have the choice of four distinct Services; to Quebec, New York, Boston, or to Philadelphia, thus enabling them to select a point of debarkation in this country nearest to their destination.

## QUEENSTOWN.

The only steamers calling at this port are those in the Liverpool and Baltimore Service. When Prepaids are issued from Queenstown our Agents there will see that the passenger is forwarded by one of the Allan Line Services most convenient to his destination.

## GALWAY.

The Allan is the only Line calling at this port, and the attention of purchasers of Prepaids whose friends reside in the West, Middle or South of Ireland, is called to the saving of time, expense and discomfort to passengers taking this route, and to the great advantage of being able to book their friends by a steamer *direct* to either Boston or Philadelphia. The steamers call at Galway regularly throughout the year, excepting the months of December, January and February. Any parties holding Prepaid Certificates from Galway and desiring to leave during those months will be furnished with free transportation to Londonderry, and passage by steamer from that port.

## NEW YORK

Is the port of landing and embarkation for passengers by Steamers in the Glasgow, Londonderry and New York Service (late State Line). Any special communications for passengers should be addressed to care of Messrs. Austin Baldwin & Co., 53 Broadway, New York.

## BOSTON.

The port of landing for passengers by steamers in the Glasgow, Londonderry, Galway and Boston Service, and is the only direct route from Scotland, the North and Middle of Ireland to the New England States. Immigrant rates to points West are the same as from New York. Company's Wharf is at Pier 5, Hoosac Tunnel Docks, Charlestown. Any special communications for passengers should be addressed to care of Messrs. H. & A. Allan, State Street, Boston.

## PHILADELPHIA

Is the port of landing for the steamers in the Glasgow, Derry, Galway, and Philadelphia Service, and is the most direct route for passengers from Scotland and Ireland, whose destination is to points in the Middle, Southern or Western States. Immigrant railroad rates from Philadelphia to Inland points in the West are considerably less than from New York. Communications for passengers expected by incoming steamers should be sent to care of H. & A. Allan, 225 Walnut St., Steamers' dock at Pennsylvania Railroad Wharf, Pier 46 South, foot of Christian Street, Philadelphia.

## BALTIMORE

Is the port of landing for steamers in the Liverpool, Queenstown, and Baltimore Service; is a direct route for passengers from British, Scandinavian and the Continental ports, to the Middle and Western States. The service is a fortnightly one, excepting about the first three months of the year. Immigrant railroad rates to Western points are generally less than those in force from New York. Any special communications for passengers should be addressed to care of Messrs. A. Schumacher & Co., 5 South Gay Street, Baltimore.

## PORTLAND, MAINE.

During the winter months the steamers in the Liverpool and Quebec Mail Service make this their terminal port, calling at Halifax to land and embark mails and passengers. The attention of New England passengers embarking at Portland should be called to the advantage of having their baggage checked to the Grand Trunk Depot, this station being adjacent to the dock.

## QUEBEC AND MONTREAL.

During the Summer months there are regular weekly Services between these ports and Liverpool, Glasgow and London. Cabin and Second Cabin passengers have the option of proceeding with the steamer to Montreal, but Steerage, as a rule, disembark at Quebec. All classes of passengers have the privilege of embarking at Montreal without additional charge. The discontinuance, during the Winter months, of the different services to the St. Lawrence is compulsory, owing to the closing of Navigation, but the Services are continued by the way of Liverpool and Portland, via Halifax, and direct from Glasgow to Halifax.

## HALIFAX, NOVA SCOTIA.

Steamers in the Liverpool and Baltimore Service call at this port, and during the Winter months the steamers in the Liverpool and Portland Service also call here to land and embark mails and passengers.

## ST. JOHN'S, NEWFOUNDLAND.

Steamers in the Liverpool, Halifax, and Baltimore Service call here fortnightly, excepting during the months of January, February, and March, when the Mail and Passenger Service between St. John's and Halifax is maintained by a special steamer.

## NORWAY, SWEDEN, DENMARK, & CONTINENT OF EUROPE.

Passengers from Scandinavian and Continental Ports are generally forwarded by the way of Hull and Liverpool to Quebec or Baltimore, and by one of the Glasgow Services when their destination is New York, Pennsylvania or New England States.

c

# PASSENGER ACCOMMODATIONS.

## CABIN.

The Dining Saloon is located on the main deck and is fitted up in an elegant and sumptuous manner. The Ladies Boudoirs and Gent's Smoking Room are situated in the best part of the ship, and are tastefully arranged, with a view to the comfort and convenience of passengers.

The State Rooms are mostly on the main deck, the majority of the rooms having two berths, and a sofa which may be used as a berth when necessary.

Rates of passage are fixed according to location of berth, all passengers having equal privileges in the saloon.

The Allan Line has always been well known for the excellence and liberality of its table, which is always provided with the delicacies of the season.

Wines and liquors of the finest qualities can be had on board at moderate prices.

All the steamers carry an experienced surgeon.

## SECOND CABIN.
### (Or Intermediate.)

This class of accommodation offers excellent opportunity to those desirous of making a trip in a most comfortable manner at an extremely small outlay.

Passengers are berthed (on day of sailing) in staterooms generally accommodating four persons, are entirely separate from the steerage, and are furnished with a liberal supply of food well cooked and neatly served—Beds, napkins, and utensils are all furnished by the Steamer.

## STEERAGE.

The Steerage, considering its cost, is remarkable for its comfort, the utmost regard having been paid to the light and ventilation of the Steamers. Male steerage passengers are berthed by themselves in one part of the ship; females and children in another.

Stewards attend on all passengers, and in addition matrons wait on females and children, in both the Second Cabin and Steerage class. Steerage passengers provide their own bedding and mess utensils, but these can be hired on board Allan Steamers at a charge of 88 cents for adults, and 44 cents for children. Food of the best quality, carefully cooked, is served by Stewards in unlimited quantity.

Competent INTERPRETERS accompany each steamer. Passengers booked through to the Western States, have also the advantages of the services of an Interpreter, who accompanies them on the train to Chicago.

## BAGGAGE.

Cabin passengers are allowed 20 cubic feet; Second Cabin 15 cubic feet, and Steerage 10 cubic feet; any excess will be charged for at the rate of one shilling sterling per cubic foot. All baggage should be labelled with name of passenger and marked "wanted" or "not wanted" as may be desired by the owner.

Passengers embarking at Montreal can forward any heavy baggage to the Company's Dock, two days prior to sailing, but it must be claimed before going aboard. Light baggage should be kept by the passengers until embarking. As large trunks cannot be taken into Stateroom, a small one, not exceeding 14 inches in height, will be found to be a great convenience, as it can be slipped under the lower berth.

## DRAFTS.

Drafts are issued for any amount at current rates by Messrs. H. & A. Allan, Montreal, on the Liverpool, London or Glasgow Offices of the Line.

# Allan Line

# Liverpool, Londonderry, Quebec and Montreal
## MAIL SERVICE.

STEAMERS SAIL EVERY WEEK DURING THE SEASON
OF ST. LAWRENCE NAVIGATION.

### CABIN PASSAGE RATES
### From Montreal or Quebec, to Londonderry or Liverpool.

## S. S. PARISIAN.

|  | | Each Passenger. | Return. |
|---|---|---|---|
| FIRST CATEGORY—Outside room for two persons...... | | $ 80 | $150 |
| SECOND " Inside room for two persons...... | | 70 | 130 |
| THIRD " Room for four persons........... | | 60 | 110 |
| Extra Class Cabins for two persons { Rooms }....... | | 100 | 185 |
| " " " for three persons { 50 to 57 }....... | | 80 | 150 |

Children between 2 and 12 years, half fare ; under 2 years, free.

Special rates for Clergymen and their families. Servants, $50 each single, $100 return.

## BY OTHER MAIL STEAMERS.

|  | | Each Passenger. | Return. |
|---|---|---|---|
| FIRST CATEGORY—Outside room for two persons...... | | $ 60 | $115 |
| SECOND " Inside room for two persons...... | | 55 | 105 |
| THIRD " Room for four persons........... | | 50 | 95 |

### BY S. S. MONGOLIAN OR NUMIDIAN.

$45 and $50, Single. $95 and $100, Return.

Children 2 to 12 years, Half Fare ; under 2 years, Free.
Special rates for Clergymen and their families.
Servants, $45 each single, $80 return.

**RETURN TICKETS** are available for passage for twelve months from date of issue.

**DEPOSIT** of $20 must be made for each berth engaged, and balance of passage money paid three days prior to sailing.

### RATES from LIVERPOOL or DERRY to QUEBEC or MONTREAL.

Cabin, $50, $70 and $80 according to accommodation.
S. S. Parisian, $60, $80 and $95.
Children 2 years and under twelve, half fare,
The rates for Cabin passage are fixed according to location of berth ; all Cabin passengers having the same saloon privileges

## SECOND CABIN RATES.

| PREPAID TO QUEBEC OR MONTREAL. | | OUTWARD FROM QUEBEC OR MONTREAL. |
|---|---|---|
| $30 | LIVERPOOL, DERRY, GLASGOW or BELFAST. | $30 |
| $33 | LONDON, BRISTOL or CARDIFF | $33 |

No reduction on Return Tickets.

Children 1 year old and under 12 half fare.

Infants, Outward free ; prepaid $3.

Second Cabin passengers are forwarded by 3rd class rail when booked beyond Liverpool, and this should be endorsed on ticket.

## STEERAGE RATES.

| PREPAID TO QUEBEC or MONTREAL. | | OUTWARD From QUEBEC or MONTREAL. |
|---|---|---|
| $20 | Liverpool, Derry, Belfast, London, Glasgow and Queenstown. | $20 |
| $21 | ............... Dublin ............... | $21 |
| $22 | ........... Bristol or Cardiff ........... | $22 |
| Half Fare | ...... Children one to twelve years ...... | Half Fare. |
| Free. | ............... Infants ............... | Free. |

Passengers booked to and from Scandinavian and Continental ports at lowest rates.

See Special Tariff for Steerage Rates to and from Inland Points in Canada.

**PASSENGER ACCOMMODATIONS UNSURPASSED.**

# RATES FROM LIVERPOOL

## To Scandinavian & Continental Points.

Only to be used in connection with First
or Second Cabin Passage.

### FROM LIVERPOOL.

| To | 1st Class. | 2nd Class. | 3rd Class. |
|---|---|---|---|
| London ............................. | $ 7 00 | $ 5 25 | $ 4 00 |
| Paris, via London, New Haven and Dieppe .................. | 15 50 | 11 50 | 8 50 |
| Paris, via London, Dover and Calais...................... | 22 00 | 16 50 | 15 50 |
| Paris, via London, Folkestone and Boulogne......... ........ | 21 00 | 16 00 | 15 00 |
| Bremen via Hull............... | 15 00 | 8 50 | 4 25 |
| Hamburg " ................. | 12 50 | 8 50 | 4 25 |
| Antwerp " ...... ......... | 9 00 | 6 00 | 4 25 |
| Amsterdam " ................. | 9 50 | ..... | 4 25 |
| Rotterdam " ................. | 9 50 | ..... | 4 25 |
| Copenhagen " ................. | 17 50 | ..... | 7 50 |
| Gothenburg " ................. | 21 00 | 14 00 | 7 50 |
| Bergen " ................. | 25 00 | 17 00 | 7 50 |
| Stavanger " ................. | 25 00 | 17 00 | 7 50 |
| Christiania " ................. | 25 00 | 17 00 | 7 50 |
| Drontheim " ................. | 41 50 | 24 50 | 7 50 |

All the above rates include Railroad fares, but not the cost of
maintenance and transfer of baggage in England. Passengers by
North Sea Steamers have to pay for their own meals on board, as
they take them.

When using these rates, Agents will please be careful and
endorse upon the Ocean Ticket the ROUTE and CLASS to which the
passenger is entitled from Liverpool.

# LIVERPOOL, QUEENSTOWN, ST. JOHN'S, HALIFAX AND BALTIMORE MAIL SERVICE.

Steamers sail regularly every fortnight throughout the year, excepting during the months of February, March and April.

## From LIVERPOOL on TUESDAYS.

## From BALTIMORE on TUESDAYS.

### RATES OF CABIN PASSAGE.

|  | Single. | Return. |
|---|---|---|
| Baltimore to Queenstown or Liverpool | $65 | $120 |
| "    to or from Halifax | 25 | 45 |
| "    to or from St. John's | 45 | 85 |
| Halifax to or from St. John's | 20 | 40 |
| "    to Queenstown or Liverpool | $60 & 65, | $110 & 120 |

Special rates for Clergymen and their families.
Halifax to Liverpool, with privilege of breaking voyage at St. John's, N.F., and proceeding to Liverpool by subsequent steamer, Cabin, $80.00.

### St. John's to or from Queenstown or Liverpool.

Cabin £15, and £13 ; Second Cabin, £8 ; Steerage, £6.
Children from 1 to 12 years, Half-fare ; under 1 year, free.

### SECOND CABIN AND STEERAGE.

|  | 2nd Cabin. | Steerage. |
|---|---|---|
| From BALTIMORE or HALIFAX to or from Liverpool, Londonderry, Queenstown, Belfast or Glasgow | $30 | $20 |
| From BALTIMORE or HALIFAX to or from London | 33 | 20 |
| Baltimore to or from Halifax | 15 | 12 |
| "    "    St. John's | 25 | 15 |
| Halifax to or from St. John's | 15 | 6 |

Children, 1 to 12 years, Half-fare.   Infants, outward, Free. Prepaid, $3.

Steamers are intended to sail at 9 A.M., on their advertised dates from Baltimore.

# ALLAN STATE } LINE.

## FIRST-CLASS PASSENGER STEAMERS.

### REGULAR WEEKLY SERVICE

BETWEEN

## New York and Glasgow

VIA

## LONDONDERRY.

THROUGH TICKETS ISSUED TO

LIVERPOOL, LONDON,

BELFAST, DUBLIN,

HAMBURG, BREMEN

AND PRINCIPAL PORTS IN

NORWAY, SWEDEN AND DENMARK.

EVERY THURSDAY
From New York.

EVERY FRIDAY
From Glasgow.

EVERY SATURDAY
From Moville (Londonderry.)

## Pier Foot of West 21st Street, New York.

# RATES SALOON PASSAGE
## New York to Glasgow via Londonderry.

### State of CALIFORNIA, NEBRASKA & NEVADA.

**S. S. STATE OF CALIFORNIA.**

| | | | | |
|---|---|---|---|---|
| Single berths in inside rooms, single trip, | $ 40 | .... | Round trip, | $ 75 |
| Inside rooms, two in a room, | " | 50 | .... | " | 9F |
| Single berths in outside rooms, | " | 45 | .... | " | 8b |
| Outside rooms, two in a room, | " | 60 | .... | " | 110 |

**S. S. STATE OF NEBRASKA.**

| | | | | |
|---|---|---|---|---|
| Forward of Saloon, outside or inside berths in the large rooms, single trip, | $ 40 | .... | Round trip, | $ 75 |
| Two in a room, inside, | " | 45 | .... | " | 85 |
| "    "    outside, | " | 50 | .... | " | 95 |
| Berths 1 to 4 and 99 to 102, | " | 45 | .... | " | 85 |
| Aft of Saloon, single berths, | " | 45 | .... | " | 85 |
| Two in a Room, | " | 60 | .... | " | 110 |

The above steamers have two large upper deck rooms capable of holding two or three persons. Prices for same can be obtained on application.

**S. S. STATE OF NEVADA.**

| | | | | |
|---|---|---|---|---|
| Berths in saloon, inside rooms, single trip, | $ 40 | .... | Round trip, | $ 75 |
| "    "    outside rooms, | " | 45 | .... | " | 85 |
| Large rooms for two persons, | " | 50 | .... | " | 95 |

Children between ages 2 and 12 yrs. half fare, Infants free.

**All Saloon Passengers have Equal Privileges.**

These steamers are lighted throughout by electricity.

**SPECIAL ADDITIONAL RAILWAY FARES.**

Glasgow to Liverpool, 3rd class.........................Single trip $2.50
"    " London    " ......................................... 5.00

Round trip tickets at double single fare.

Passengers have their choice of lines, and are allowed 10 days stop-over at Glasgow if desired.

**No Cattle are Carried on these Steamers.**

Holders of Round Trip or Excursion Tickets have the option of returning by the ALLAN LINE ROYAL MAIL STEAMERS, sailing from Liverpool to Montreal during the summer season, on presentation of their certificates to Messrs. ALLAN BROS. & Co., James Street, Liverpool. Additional fare will be charged by this service according to the steamer and location of stateroom.

### SECOND CABIN.

From New York to Glasgow, Derry, Belfast or Liverpool.............. $ 30
From Glasgow, Derry, Belfast or Liverpool to New York............. 30
Round Trip........................................................ 55
Children between 1 and 12 years, half fare. Infants, Outward, Free.
    Infants, Prepaid............................................. 3.00

# RATES OF STEERAGE PASSAGE.

### *New York and Glasgow Service.*

#### OUTWARD FROM NEW YORK.

| | |
|---|---|
| To Glasgow, Londonderry, Liverpool or Belfast .......... | $20 |
| To Dublin.................................................. | 21 |
| To London, Bristol or Cardiff............................. | 22 |

##### ALL RAIL FROM GLASGOW OR LONDONDERRY.

| | |
|---|---|
| To Copenhagen, Malmo, Gothenburg, Christiania, Christiansand, Stavanger, Bergen, Throndhjem and Esbjerg................. | 24 |
| To Stockholm.............................................. | 27 |
| To Hamburg, Bremen, Amsterdam, Rotterdam or Antwerp......... | 22 |

Children between 1 and 12 years of age, half fare.

Infants under 1 year, { To British Ports, ...................... Free.
{ To Scandinavian Ports................. 3.00

#### PREPAID TO NEW YORK.

| | |
|---|---|
| From Glasgow, Londonderry, Belfast or Liverpool ........ | $ 20 |
| From Dublin.............................................. | 21 |
| " London, Bristol or Cardiff............................ | 22 |
| " Hamburg, Bremen, Amsterdam, Antwerp or Rotterdam....... | 23 |
| " Copenhagen, Malmo, Gothenburg, Christiania, Christiansand, Stavanger, Bergen, Throndhjem and Esbjerg................ | 24 |
| " Stockholm............................................. | 27 |

Children between 1 and 12 years of age, half fare. Infants,....... $3.00

### NOTE.

Passengers booked to or from Scandinavian or German Ports are forwarded by Leith, Hull or Newcastle, making close connections with our steamers. Liberal quantity of provisions well cooked will be served three times a day by the Company's stewards.

An experienced surgeon with the steerage compartments; medical attendance and medicine free of charge.

Separate compartments are assigned to single men, single women and married couples.

# FORWARDING RATES.

### *New York and Glasgow Service.*

Additional Fares for booking through from Glasgow to London and the continent.

## FROM GLASGOW.

|  | 1st Class. | 2nd Class. | 3rd Class. |
|---|---|---|---|
| To Liverpool, ............... | $ 8.00 | $ 5.80 | $ 2.50 |
| " London ................. | 14.00 | 10.00 | 5.00 |
| Londonderry to Dublin,....... | — | — | 1.85 |
| " Belfast........ | — | — | 1.30 |
| To Christiansand Copenhagen, } via Leith.. | 16.00 | — | 8.50 |
| " Christiania, Stavanger, } via Hull .... | 26.50 | 18.50 | — |
| " Bergen, via Newcastle, ..... | 17.50 | — | — |
| " Gothenburg, via Hull, ..... | 22.50 | 16.00 | — |
| " Hamburg, Bremen, } via Leith,...... | 15.00 | — | 7.00 |

All the above rates include Railroad fares, but not the cost of maintenance and transfer of baggage in England. Passengers by North Sea Steamers have to pay for their own meals on board as they take them.

When using these rates, Agents will please be careful and endorse upon the Ocean Ticket the ROUTE and CLASS to which the passenger is entitled from Glasgow.

# London, Quebec and Montreal Service.

## STEAMERS SAIL EVERY FORTNIGHT.

### Rates of Passage from London to Quebec or Montreal.

Cabin..................................................$50 00
Second Cabin......................................... 30 00
Steerage ............................................ 20 00

Children 1 to 12 years, half-fare.

These Steamers do not carry passengers on the voyage
to London.

# Glasgow, Quebec and Montreal Service.

## STEAMERS SAIL REGULARLY EVERY WEEK.

### Rates of Passage from Glasgow to Quebec or Montreal.

Cabin..................................................$50 00
Second Cabin......................................... 30 00
Steerage ............................................ 20 00

Children 1 to 12 years, half fare.

Steamers in this Service do not carry passengers on the
voyage to Glasgow.

Persons sending for their friends from Scotland, or the South
of England, should book them by one of the above DIRECT routes.

# GLASGOW AND BOSTON SERVICE,

## Via LONDONDERRY & GALWAY.

---

### Steamers are Intended to Sail Every Fortnight.

---

## THE ONLY DIRECT ROUTE

# TO BOSTON,

### AND THE NEW ENGLAND STATES,

—FROM—

# GLASGOW,

# LONDONDERRY and

# GALWAY.

---

## RATES OF PASSAGE TO BOSTON.

|  | 2nd Cabin. | Steerage. |
|---|---|---|
| From Glasgow, Liverpool, Derry, Belfast or Galway............................... | $30 | $19 |
| From London.............................. | 33 | 20 |
| " Bristol or Cardiff...................... | 33 | 21 |

Children 1 to 12 years, half fare ; Infants, $3.

Passengers booked from Continental and Scandinavian points at lowest rates.

Steamers in this Service do not carry passengers FROM Boston.

# GLASGOW AND PHILADELPHIA SERVICE,

### Via LONDONDERRY AND GALWAY.

## *EVERY FORTNIGHT.*

## THE ONLY DIRECT ROUTE,

# TO PHILADELPHIA,

### And the MIDDLE, SOUTHERN and WESTERN STATES,

—FROM—

## GLASGOW, LONDONDERRY AND GALWAY.

### *RATES OF PASSAGE TO PHILADELPHIA.*

|  | 2nd Cabin. | Steerage. |
|---|---|---|
| From Glasgow, Liverpool, Derry, Belfast or Glasgow .................................. | $30 | $19 |
| From London................................ | 33 | 20 |
| " Bristol or Cardiff...................... | 33 | 21 |

Children 1 to 12 years, half fare ; Infants, $3.

Passengers booked from Continental and Scandinavian points at lowest rates.

The attention of parties sending for their friends from Scotland, or the North and Middle of Ireland, is particularly called to the convenience of the above route. Passengers are saved all the annoyances attending transfer to Liverpool, and have the advantage of being landed at an American port nearest their destination.

Steamers in the above service do not carry passengers on the homeward voyage to Glasgow.

# PREPAID CERTIFICATES

### ISSUED FROM

## LIVERPOOL, LONDON,
## QUEENSTOWN, GALWAY,
## GLASGOW, LONDONDERRY,
## BELFAST,
## GERMANY, NORWAY,
## SWEDEN,
## DENMARK, FINLAND.

PREPAID CERTIFICATES are issued to persons who desire to pay on this side of the Atlantic for the passage of friends or relatives from Europe. Passengers can be booked from almost any Railroad Station in Great Britain and Ireland, or the principal Continental and Scandinavian ports through to their destination in the United States or Canada, by adding amount of railroad fare to the Ocean passage.

If you purpose bringing out your friends from the old country, do not make arrangements without first calling upon the Agent of the **ALLAN LINE** in your City, so that he may advise with you regarding the most **DIRECT ROUTE.**

It should be borne in mind that in many instances, passengers by not taking a direct steamer, suffer more inconvenience through the annoyances attending transfer, than during the entire Ocean voyage.

Certificates are good for passage for twelve months from date of issue, and if not used within that time, can be extended for a further period on payment of any advance in rates.

### Rates always as low as by any other first class Line.

---

**(partial text from adjacent page, left margin)**

RVICE,

r.

TE,

,

STATES,

VAY.

Steerage.

$19
20
21

1 points

otland,
to the
all the
antage
n.

on the

# PRINCIPAL AGENCIES.

LIVERPOOL......................Allan Brothers & Co., James Street.
PLYMOUTH......................Weekes, Phillips & Co.
LONDON............ { Allan Brothers & Co., 103 Leadenhall Street.
                  { Montgomerie & Workman, 96 Gracechurch Street.
NEWCASTLE.........................Pynan Bell & Co.
CARDIFF............................Geo. Bird, 246 Bute Street.
BRISTOL...........................John W. Down, Bath Bridge.
LONDONDERRY.................Allan Brothers & Co., Foyle Street.
GALWAY............................J. & A. Allan.
QUEENSTOWN......................James Scott & Co.
DUBLIN.....................Thos. O'Dell & Co., 12 Eden Quay.
BELFAST.......................A. Thomson & Co., 14 Victoria Street.
GLASGOW.......................J. & A. Allan, 25 Bothwell Street.
PARIS................Pitt & Scott, 7 Rue Scribe, Passenger Agents.
MARSEILLES......T. L. Louis, 106 Boulevard des Dames, Pass'gr. Agent.
ANTWERP.............. } Apply to
GRONINGEN............ } Richard Berns, 132 Avenue de Commerce
HARLINGEN............ } Antwerp.
ROTTERDAM.........H. G. Botermans, Wynstraat 116, Pass. Agent.
AMSTERDAM.................Bruinnier & Co., Passenger Agents.
HAMBURG......B. Karlsberg, Admiralitatstrasse, 31, Agent N. Y. Service.
HAMBURG, STETTIN, } Apply to Spiro & Co.,
BERLIN, BREMEN.    } 6 Bahnhof Strasse, Hamburg.
BREMEN........................Harry Cohen, Agt. N. Y. Service.
MANNHEIM...............Libra F. 7. No. 31 Gebruder Bielsfeld.
COPENHAGEN............Albert Moller, Nyhavn 13, Passenger Agent.
MALMO.......................N. P. Londahl, Ostra Hamnkaien.
HELSINGBORG..............A. Andersson, Norra Strandgaten 13.
GOTHENBURG......................J. Freder, Passenger Agent.
GOTHENBURG.............C. J. Edenholme, Agent N. Y. Service.
CHRISTIANIA....................A. Sharpe, Passenger Agent.
BERGEN...................Johan Martens, Passenger Agent.
CHRISTIANSAND.............C. F. C. Ullitz, Passenger Agent.
STAVANGER..................J. L. Walthoe, Passenger Agent.
DRONTHEIM....................Joh. Mathison, 1 Braterveiten.
STOCKHOLM................Allan Liniens, Afdelningskontor.
HANGO, FINLAND..........H. Boostrom, Finska Angfartygs Co.
ULEABORG..............W. Otto Rovander, Finska Angfartygs Co.
WASA............... { Cairenius & Co., Finska Angfartygs Co.
OSTERMYRA......... {
ABO....................Ferdinand Frenkell, Finska Angfartygs Co.
HELSINGFORS, FINLAND.........Lars Krogius, Finska Angfartygs Co.
QUEBEC..........................Allans, Rae & Co.
CHICAGO...................Allan & Co., 112 La Salle Street.
HALIFAX, N.S.......................S. Cunard & Co.
BALTIMORE.......................A. Schumacher & Co.
ST. JOHNS, N.F......................Shea & Co.
ST. JOHN N.B....................Wm. Thomson & Co.
CHARLOTTETOWN, P.E.I.................Carvell Bros.
TORONTO..........................H. Bourlier.
**NEW YORK.....Austin Baldwin & Co., 53 Broadway.**

### H. & A. ALLAN, GENERAL AGENTS,

BOSTON.            PHILADELPHIA.            MONTREAL.

STATE OF CALIFORNIA.　　　STATE OF NEVADA.
STATE OF NEBRASKA.　　　STATE OF PENNSYLVANIA.

Printed in the USA
CPSIA information can be obtained
at www.ICGtesting.com
LVHW012329151023
761173LV00012B/800